DATE DUE

OCT 2 2 2001			
GAYLORD 234			PRINTED IN U. S. A.

One for a penny.
Ten for a dime.
Count them and buy them.
You'll have a good time!

  1 penny = 1¢

 1 nickel = 5¢

 1 dime = 10¢

 1 quarter = 25¢

JELLY BEANS FOR SALE

Written and photo-illustrated by **Bruce McMillan**

SCHOLASTIC PRESS ● New York

1¢ = 1 jelly bean

1¢ + 1¢ + 1¢ + 1¢ + 1¢ = 5 jelly beans

5¢ = 5 jelly beans

$1¢ + 1¢ + 1¢ + 1¢ + 1¢ + 5¢ = $ **10 jelly beans**

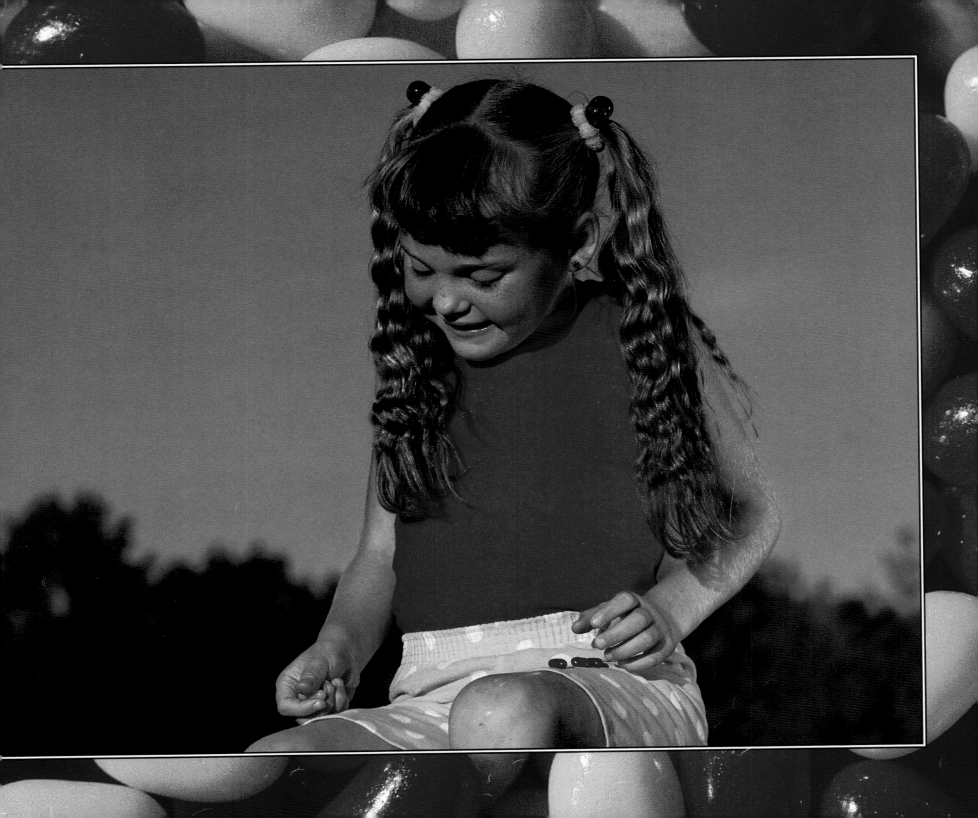

5¢ + 5¢ = 10 jelly beans

10¢ = 10 jelly beans

1¢ + 1¢ + 1¢ + 1¢ + 1¢ + 1¢ + 1¢ + 1¢ + 1¢ + 1¢ = 10 jelly beans

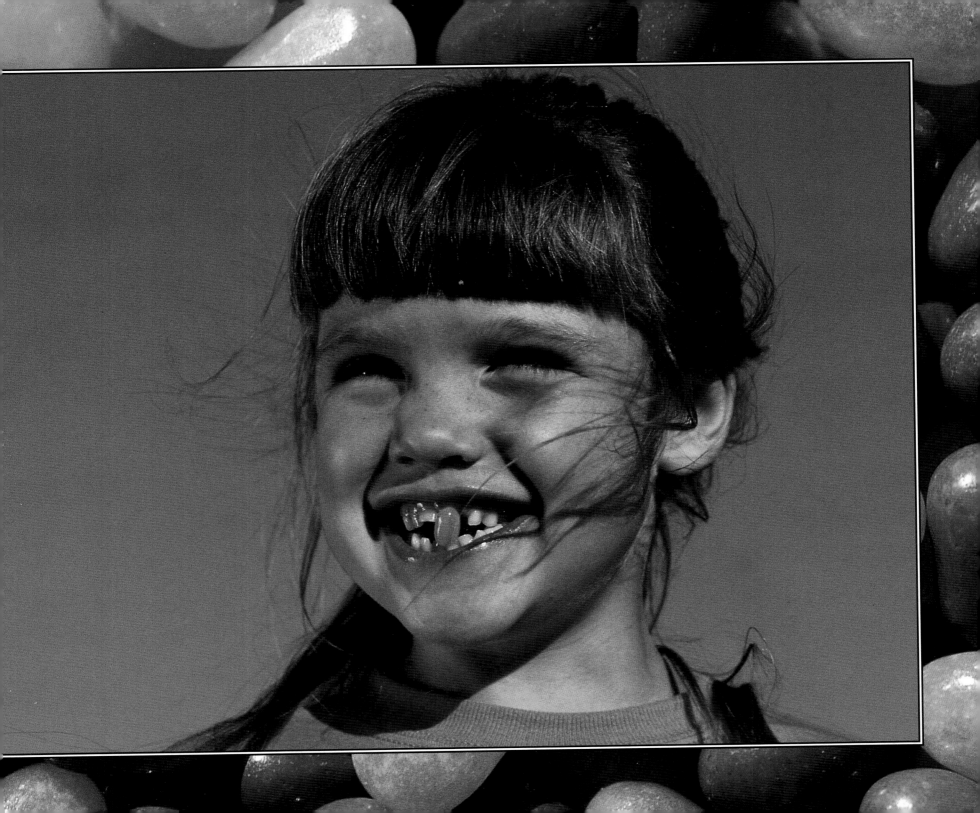

5¢ + 5¢ + 5¢ + 1¢ + 1¢ + 1¢ + 1¢ + 1¢ + 1¢ + 1¢ + 1¢ + 1¢ + 1¢ = 25 jelly beans

$$1¢ + 1¢ + 1¢ + 1¢ + 1¢ + 5¢ + 5¢ + 10¢ \ = \ 25 \text{ jelly beans}$$

10¢ + 10¢ + 1¢ + 1¢ + 1¢ + 1¢ + 1¢ = 25 jelly beans

10¢ + 10¢ + 5¢ = 25 jelly beans

1¢ + 1¢ + 1¢ + 1¢ + 1¢ + 1¢ + 1¢ + 1¢ + 1¢ + 1¢ + 1¢ + 1¢ +
1¢ + 1¢ + 1¢ + 1¢ + 1¢ + 1¢ + 1¢ + 1¢ + 1¢ + 1¢ + 1¢ + 1¢ + 1¢   =   25 jelly beans

25¢ = 25 jelly beans

Colors   Flavors

Colors	Flavors
red	cinnamon
orange	orange sherbet
yellow	sour lemon
green	jalapeño
blue	blueberry
purple	island punch
brown	cookies 'n' cream
white	coconut
pink	cotton candy

The History of Jelly Beans

Jelly beans were invented and named in the United States during the 1800s. Yet their two-part manufacturing process has international roots. The chewy gel centers of jelly beans are descendants of Turkish delight, a fruit-jelly candy that has been made in the Middle East for hundreds of years. Their outer shells are descendants of Jordan almonds. They are made using a process called panning, invented in France during the 1600s. A pan filled with almonds, sugar, and syrup is rocked until the almonds are coated with a hard, sweet covering.

Making These Jelly Beans

Forty million jelly beans can roll off the production line in a day. But to make them, it takes seven to ten days from start to finish.

Day One: To make the centers, a liquid syrup of water, cornstarch, sugar, and corn syrup is mixed together. Flavors are added, such as the purees and concentrates of blueberries, lemons, or even jalapeño peppers. The liquid mixture is cooked. Then powdery cornstarch is packed into shallow trays and impressed with jelly bean shapes (like footprints on damp beach sand). The liquid mixture is dripped into each tiny impression, then the trays are put into a high-temperature, low-humidity room overnight, where the centers begin to harden and take their bean shapes.

Day Two: The centers are turned out of the molds and shaken to remove the cornstarch. Then they get a steam bath, and to keep them from sticking to one another, they get a sugar shower.

Day Three: The centers get a day to rest and harden a little more.

Day Four: Now it's time for panning—the shell-making process. The centers are put into a kettledrum-shaped copper pan. Its inside surface is covered with indentations. It rotates like a cement mixer, and for two hours the drum slowly turns. Flavored and colored syrups are hand poured over the beans four times, coating them evenly as they tumble.

Day Five: To add a glossy finish, the beans are placed in a rounded, stainless steel pan, and as they tumble, they get coated with natural waxes. A polishing of confectioner's glaze completes the process.

Day Six: The still-sticky beans rest until the shells are completely set.

Day Seven: The beans are ready to pack. However, some flavors that use fruit juice and citric acid need up to an extra three days of resting during the process so they won't stick to one another.

Presidents, Coins, and Jelly Beans

United States presidents are portrayed on the fronts of U.S. coins. In fact, President Lincoln is on both sides of the penny. If you look closely on the back, you'll see him sitting in the center of the Lincoln Memorial, looking right at you! But it was a president who is not on a coin, President Reagan, who helped make famous the brand of jelly beans photographed in this book. In 1980, the Jelly Belly blueberry jelly bean was invented specifically to honor Mr. Reagan on his first inauguration. Three-and-a-half tons of red, white, and blue jelly beans were sent to Washington, D.C. As president, he always kept a bowl of this brand of jelly beans in the White House's Oval Office, and passed them around at cabinet meetings. President Reagan also arranged to have them sent into space as a snack for the astronauts. So they were the first jelly beans launched into outer space, and they orbited Earth, June 18 to 24, 1983, aboard the space shuttle *Challenger*. That also means they were the first jelly beans to float in space as the astronauts ate them!

A Note from the Author

The children who ran our jelly bean stand in Shapleigh, Maine, are Arianna Obertautsch and Ross April. Their customers are, in order of appearance, Jordan Chadbourne, Priscilla Arsenault, Panhar Thoim, Kristen Hodge, Jonathan Barber, Joshua Barber, Maxwell Schnur, Marie Frizzell, Wesley Gowen, Andrea Fajardo, Fidel Tham, Kalie Snow, Billy Barry, and Paula Tham. I thank not only them, but especially their parents for helping me match the color of the children's clothes to the jelly beans, and for helping during the shoot. I also thank Eleanor Loija for the use of her field.

I used a Nikon F / MF23 with a 105 mm micro lens, sometimes with a polarizing or light blue filter. I shot with full sunlight in the late afternoon and early evening, and used Kodachrome 64 film, which was processed by Kodak. The coins were photographed on the part of the stand that I made of red cedar.

Free Jelly Bean Kits for Classrooms

The jelly beans photographed in this book are Jelly Belly gourmet jelly beans. (They were provided by the Herman Goelitz Candy Company, Inc., 2400 North Watney Way, Fairfield, CA 94533-6741, through the courtesy of Peter Cain, Vice President of Marketing.) The following is subject to change: Elementary teachers and schools may obtain a free classroom kit containing an eight-minute factory tour video, a half-pound bag of jelly beans, a teacher's guide, a poster, stickers, pamphlets, and menus by writing or faxing toll free to: Video Placement Worldwide, P.O. Box 58142, St. Petersburg, FL 33715-9976, fax 1-800-358-5218. For more information, call the Jelly Belly hotline toll free at 1-800-522-3267.

For Dianne

Library of Congress Cataloging-in-Publication Data
McMillan, Bruce
Jelly beans for sale / written and photo-illustrated by Bruce McMillan. p. cm.
Summary: Shows how different combinations of pennies, nickels, dimes, and quarters
can buy varying amounts of jelly beans. Includes information on how jelly beans are made.
ISBN 0-590-86584-6
1. Addition—Juvenile literature.   2. Coinage—United States—
Juvenile literature.   3. Jelly beans—Juvenile literature.
[1. Addition. 2. Coins   3. Money. 4. Jelly beans.] I. Title.
QA115.M395   1996   332.4'04—dc20   95-25864   CIP   AC
12 11 10 9 8 7 6 5 4   7 8 9/9 0 1/0
Printed in the U.S.A.   36
First edition, September 1996

The text type was set in Optima.
Design by Bruce McMillan